CW01665700

RAIL 13 PORTFOLIOS

The AC Electrics

Compiled by Chris Shaw

First published 1991

ISBN 0 7110 1938 X

All rights reserved. No part of this book may be reproduced or transmitted in any form or by any means, electronic or mechanical, including photocopying, recording or by any information storage and retrieval system, without permission from the Publisher in writing.

© Ian Allan Ltd 1991

Published by

IAN ALLAN LTD

Terminal House Shepperton TW17 8AS

Telephone: Walton-on-Thames (0932) 228950

Fax: 0932 232366 Telex: 929806 IALLAN G

Registered Office: Terminal House Shepperton TW17 8AS

Printed by Ian Allan Printing Ltd at its works at Coombelands in Runnymede England

Front cover:
With the British Steel Corporation's Hardendale Works towering in the background, Nos 87010 *King Arthur* and 86436 cautiously propel their train of PTA empties into the quarry sidings on the morning of 2 December 1987. *Dave McAlone Bronica ETRS 150mm Zenzanon Ektachrome 200 1/250, f11*

Back cover:
'AL5' No E3056 climbs out of Runcorn with the 16.15 Liverpool Lime Street/Southport-Euston on 20 July 1963. *Michael Mensing*
Voigtlander Bessa II 105mm Color-Heliar High-speed Ektachrome 1/500, f6.3

Introduction

For some reason the railway enthusiast world has been slow to recognise the attraction of British Rail's ac electric locomotive fleet. It should not be forgotten that when the first production locomotive, No E3001, was rolled out at Crewe late in 1959 it created a considerable stir. For here, gleaming in a highly modernist semi-metallic blue livery and with smartly raked-back nose ends, was a clear vision of the future. Forget the Pilot Scheme diesels that had emerged by that time; they seemed cumbersome by comparison. It was the electrics, and the improvements in infrastructure that were made for them, that fuelled the imagination of BR's commercial managers, faced as they were with the first real signs of the postwar boom in airline and road competition over routes hitherto the railways' almost exclusive preserve.

Government indecision over the funding of electrification schemes meant that it took a full six years for electrification to reach from Manchester, Liverpool and Crewe southwards to Euston, by which time the second generation 'AL6' design was being introduced. Of rather bland appearance compared with the initial 100 locomotives, the 'AL6s' consolidated the ac electric fleet, providing slightly more power and considerably improved reliability in service. A total of 100 locomotives was built, making it by far the largest class numerically and this, together with the design's outstanding competence, has resulted in the 'AL6s' performing the lion's share of ac electric work over the years. Indeed, so successful was the class that when an order for 35 (later 36) locomotives was placed with GEC in 1971 for the Glasgow extension the resulting Class 87 was, in effect, an uprated and improved version of the earlier design, and with it BR possessed a traction unit that could take the infamous gradients on the northern fells on the West Coast main line easily in its stride.

Interest in the ac fleet was stimulated in 1984 by the application of two experimental liveries to members of Class 87, from which were developed the current InterCity and main line paint schemes. One locomotive, No 86401, later received Network SouthEast livery, and in 1988 the short-lived Freightliner Class 86/5s were painted in Railfreight livery, thereby creating a further variant, perpetuated on Railfreight's subsequent Class 86/6 conversions and Class 90s. Yet another new electric livery was unveiled in July 1990 when Parcels-allocated No 86419 was outshopped from Stratford in that Sector's new red and grey colour scheme.

With the latest Class 90 and 91 locomotives now demonstrating their impressive capabilities in traffic, the puissant Class 89 waiting in the wings for one of the business sectors to grasp the nettle, and orders for the Class 92 general freight locomotive design placed with Brush, the ac electric scene has probably never been so full of interest.

It has been my intention in compiling this volume to present a representative overview of ac locomotive activity during the last 30 years or so, and I offer sincere thanks to the many photographers who so kindly made their collections available for publication. I would also like to thank Roger Wood for his help in checking the original manuscript. Particular gratitude is due to those photographers active in the 1960s who were able, just occasionally, to turn their backs on the decline of steam and record the dawning of a new era; without their efforts we would be perceptibly the poorer.

CHRIS SHAW
Shepperton
July 1990

Previous page:
One of the most spectacular locomotives to run on BR for many years is Class 89 No 89001 *Avocet*. Built at BREL Crewe to a Brush design this highly successful 105-ton, 5,800hp prototype has amassed a considerable mileage on both the West Coast and East Coast main lines. The wide-angle lens captured the locomotive at Brush Electrical Machines' Loughborough works on 12 August 1989 awaiting the addition of Time Division Multiplex (TDM) equipment for push-pull working. *Brian Morrison*
Nikon FE2 24mm Nikkor Ektachrome 64 1/125, f8

Right:
Barely six months old when this picture was taken on 22 September 1961, 'AL1' (later Class 81) No E3020 arrives at Crewe with a van train off the Manchester line. The locomotive was based at Crewe Electric Traction Depot (ETD) at this time, but the 'AL1s' were later to become associated with Scotland when in March 1975 the whole class was reallocated to Glasgow Shields Road to provide secondary motive power for the newly-commissioned Anglo-Scottish electrification. *W. Potter*
Kodak Retinette IIB Kodachrome II 1/250, f4

E 3020
E 3020

Left:
'AL2' No E3046 and 'AL3' No E3027 pass through Crewe station on their way to Crewe ETD, having worked a train in from Manchester Piccadilly. The use of electric locomotives for the 31-mile section between Manchester and Crewe may seem to have been wasteful in hindsight, but it did provide the opportunity thoroughly to commission the fleet and gain service experience pending electrification southwards to Rugby and Euston. *W. Potter*
Kodak Retinette IIB Perutz 40ASA 1/250, f3.5

Left:
Arguably the most elegant and neatly styled of the early series ac electrics, the 'AL4s' (later Class 84s) sadly were the least reliable in service, failures due to flashovers and wear on drive components resulting in their early demise. In their latter years the locomotives acquired something of a cult following amongst enthusiasts arising from their rarity on Class 1 work, and they were rostered power on several railtours.

This picture shows No E3042 (No 84007) slowing for the Crewe stop with the 11.10 Manchester-Cardiff/Paignton on 5 March 1961. Note the generator van marshalled behind the locomotive to provide steam heating for the train.
Michael Mensing
Retina IIA 50mm Xenon Agfa CT18 1/500, f4

Right:
'AL3' No E3030 races towards Hademore crossing, south of Lichfield, with the 13.30 Euston-Perth on 25 July 1964. Many will best remember the 'AL3s' for the loud whine made by the traction motor cooling fans when the locomotives were at rest, a characteristic shared with the other early classes, albeit to a lesser extent, and which gave rise to the 'roarer' nickname by which Types AL1-AL5 are affectionately known to enthusiasts.
Michael Mensing
Nikkorex F 50mm Nikkor Agfa CT18
1/1000, f2.2

Right:
By November 1964 electrification had reached as far south as Rugby where a changeover from electric to diesel traction remained necessary until the scheme was commissioned through to Euston in 1966. AEI's Rugby works is an appropriate backdrop as 'AL1' No E3017 and 'AL5' No E3089 arrive with the Up 'Ulster Express' on the cold, crisp morning of 28 December 1964. *Bryan Hicks*
Agfa Silette Agfa CT18 1/300, f4.5

Crossing the Warwickshire/Northamptonshire border near Kilsby is 'AL5' No E3073 with an up express on the afternoon of 16 April 1967. The previous April had seen the introduction of an accelerated timetable for services to Manchester and Liverpool based on full electric working to and from Euston, and it had been intended that all the coaches for this service would be painted in the then-new blue and grey livery. Although this was achieved, maroon-liveried vehicles continued to appear in other West Coast main line sets for some time to come, doubtless much to the chagrin of the London Midland's marketing people!
Bryan Hicks Baldamatic I Agfa CT18 1/125, f8

Left:
Leaving Market Harborough flyover, east of Rugby station, is 'AL2' No E3055 crossing from the Up to the Down side of the station layout on 22 May 1965. Electrification of the now closed Market Harborough line as far as Clifton Mill gave this useful facility which was of particular value when Rugby was the limit of electric working and frequent run-round movements were necessary.

No E3055 was unique among the first 100 locomotives in being built with two AEI crossarm pantographs (later removed) instead of the Stone Faiveley type fitted to the other early electrics. Sadly this locomotive was destined for a short life, being irreparably damaged by fire in 1966. It was put into store while consideration was given to rebuilding as a thyristor prototype, possibly with AEI sponsorship, but nothing positive resulted and official withdrawal followed in September 1969, scrapping taking place at BREL Crewe the following August. *Michael Mensing*
Nikkorex F 50mm Nikkor Agfa CT18 1/500, f3.5

Left:
Willesden District Electric Depot was opened in 1965 to service the new 100-strong 'AL6' fleet introduced to coincide with the start of full electric services to London. 'AL6' construction was divided between BR at Doncaster, which built 40 locomotives, and English Electric at Vulcan Foundry which produced the remainder. Livery was an early version of the new rail blue, although the yellow warning panel was a later addition. This picture shows one of the English Electric examples, No E3169, receiving attention at Willesden on 7 October 1967. *John Glover*
Ilford Sportsman Agfa CT18 1/125, f5.6

Right:
The tranquility of Blisworth, a few miles south of Rugby, is disturbed as 'AL1' No E3002 rushes past with a Down North Wales express on 29 May 1966. To a large number of enthusiasts at the time the ac electrics represented the absolute nadir of the modern railway and a tangible result of Dr Beeching's chairmanship, but the travelling public welcomed the spectacular acceleration that the locomotives brought to their services. The modern era had arrived with a vengeance! *Bryan Hicks*
Baldamatic 1 Agfa CT18 1/500, f4

Right:
Unfortunately No E3002 was to enjoy less than nine years' service, being damaged by a traction motor fire and withdrawn in November 1968. The locomotive was moved to BREL Crewe where it is seen in the process of being cut up. *John Glover*
Praktica Nova 50mm Tessar Agfa CT18
1/125, f5.6

In order to convert the ac supply from the overhead line to dc for the traction motors the ac electrics were provided with rectifiers, Types AL1-AL4 having a mercury-arc type while the 'AL5s' introduced a relatively unproven germanium design. The mercury-arc rectifiers proved to be extremely unreliable in service, particularly those fitted in Types AL3 and AL4, and various attempts were made to find a solution. The situation became so critical that BR was even considering withdrawal of these classes, but completion of 'AL6' deliveries in the summer of 1966 permitted indefinite storage of the two types without jeopardising services. The rather unusual, but secure, location chosen was Bury steam shed and the first locomotives found themselves there from late 1967. The remainder of Types AL3 and AL4 followed with the exception of No E3043 which went to Rugby Testing Station for rectifier experiments to be carried out. Occasionally being wheeled out for exhibitions of the latest in BR traction, an irony if ever there was one, the announcement of the Glasgow electrification scheme proved to be the saviour of the two classes. All four of the mercury-arc-equipped types were converted to silicon diode rectification and by 1972 the whole of the first generation ac electric fleet was back in traffic.

This rare photograph, taken from a passing EMU, shows four 'AL3s' and an 'AL4' standing disconsolately at Bury on 11 February 1970.
John Glover
Miranda Automex III 50mm Auto-Miranda
Agfa CT18 1/250, f5.6

Euston
1A18
E3122

Left:
The accelerated West Coast main line electric timetable introduced on 18 April 1967 featured the first daily Pullman trains ever to appear at Euston, comprising a Manchester all-Pullman service and a Liverpool service which also conveyed second-class stock. New cars, built to Mk 2 specification, were provided and were distinctive in carrying a reversed version of the new standard blue and grey livery. 'AL6' No E3122 hums into the rebuilt London terminus with the Up morning 'Liverpool Pullman' on 12 May 1967.
Euslin Bruce/J. Salmon Ltd, Sevenoaks, Kent
Wirgin 105mm Schneidar-Kreuznach Radionar
High-speed Ektachrome

Above:
'AL5' No E3060 (later No 85005) pauses at Stafford with an up express in the mid-1960s. The locomotive is still carrying 'electric blue' livery, although small yellow warning panels have been added and the second pantograph has been removed. *Norman Preedy*
Ensign Selfix 820 High-speed Ektachrome
1/100, f8

Stafford again, this time on 10 June 1969 with 'AL5' No E3075 passing through the station light engine. Comparison with the previous photograph reveals the loss of style engendered by the repainting of the early locomotives in corporate blue livery.

Types AL1-AL5 developed something of a reputation for poor riding, and complaints by drivers were frequent; indeed on at least one occasion a 'wet patch' in the track caused such a jolt that a driver was thrown out of his seat! Tests were conducted, the result of which was to alter bogie maintenance schedules in an attempt to optimise performance. Nevertheless, the riding of these locomotives remains notorious to this day.

John Glover

Miranda Automex III 50mm Auto-Miranda

Agfa CT18 1/125, f8

Left:
A Down fitted express freight takes the Trent Valley line at Rugby in the care of 'AL6' No E3155 on 7 March 1969. Although this design was based closely on the previous 'AL5', it incorporated more modern styling without the raked-back nose ends that characterised the early classes. A departure from existing practice was the adoption of axle-hung nose-suspended motors, a decision dictated primarily by considerations of cost and maintenance. This configuration had been successfully used in diesel locomotives, but it soon became apparent that the forces exerted on the track by the resulting unsprung weight, especially at sustained high speeds, were having an adverse effect, not only on the permanent way, but on drivers' backsides as well! *Bryan Hicks*
Pentax S1A Agfa CT18 1/500, f4

Left:
Electrification of the route from Rugby to Stafford via the West Midlands was completed late in 1966 when the Wolverhampton-Birmingham-Rugby section was energised on 6 December. The work included modernisation of many of the stations. Stechford, a particularly soulless example of 1960s rebuilding, is the location as 'AL3' No E3029 decelerates for the New Street stop with the 12.15 Euston-Manchester Piccadilly on 12 August 1968.
Michael Mensing
Nikkorex F 50mm Nikkor Agfa CT18 1/500, f2.8

Storming along the Trent Valley main line at Easenhall, between Rugby and Nuneaton, is 'AL3' No E3035 (later No 83012) with the 14.05 Euston-Glasgow on 24 August 1968. Introduced in July 1961, this locomotive was to endure two periods of storage during its life; at Bury from March 1969 to February 1972 (see page 11) and at Longsight from November 1982. In 1983, however, the surprise decision was taken to reinstate it to traffic, together with No 83015 and two members of Class 82, to work empty coaching stock trains between Euston and Stonebridge Park carriage sidings. Although restricted to 40mph for this unglamorous work the locomotive gained quite a following, no doubt helped by a repaint into InterCity Sector livery in 1986. Sadly March 1989 saw the withdrawal of No 83012, and with it Classes 82 and 83, after a life of over 28 years. *Michael Mensing*
Nikkorex F 50mm Nikkor Agfa CT18 1/1000, f2.2

Representing a typical West Coast main line express of the late 1960s/early 1970s period 'AL6' No E3196 (later No 86219 *Phoenix*) is seen south of Polesworth with the 12.45 Holyhead-Euston on 18 August 1968. The formation comprises a mixture of Mk 1 and early Mk 2 vehicles. *Michael Mensing*
Nikkorex F 50mm Nikkor Agfa CT18
1/1000, f2.2

Right:
The characteristic pink flowers of rose bay willow-herb contribute to the atmosphere of neglect as No E2001 nears the end of its somewhat obscure life, dumped in Rugby EMU sidings on 5 August 1972. The black and silver livery has succumbed to the rigours of rust and the elements and the locomotive's original number, E18100, can just be discerned showing through the grime. It was broken up by J. Cashmore of Tipton in January the following year. *Peter Shoesmith*

Right:
More willowherb is seen lining the embankment just north of Kilsby Tunnel as Class 86/0 No 86018 (formerly 'AL6' No E3163) emerges with the 15.45 Euston-Glasgow on 2 August 1974. As originally built Class 86 was divided into two distinct varieties; those with GEC 282AZ traction motors, giving a continuous rating of 3,600hp in total, and those with the 282BZ type, giving 4,040hp, although initially no distinction was made between these for traffic purposes. *Terry Flinders*
Pentax Spotmatic 50mm Takumar
Ektachrome X 1/250, f5.6-8

Concurrent with electrification to the West Midlands was a scheme to wire the 'alternative' route to Manchester Piccadilly from the junction at Colwich, south of Stafford, via Stoke-on-Trent. This section, together with the 3½-mile link from Norton Bridge to Stone, was energised on 5 December 1966. Curiously the opportunity to electrify the 8¾-mile line from Crewe to Kidsgrove was not taken, thereby depriving the operators of a potentially useful electrified diversionary route. The scene is Stoke-on-Trent, with a clean Class 86, No E3183, drawing a rake of air-conditioned Mk 2 stock out of the station forming a Manchester Piccadilly-Euston service on 7 August 1973. On the left Class 47 diesel-electric No 1830, still in green livery, awaits a path with a merry-go-round coal train.
Hugh Ballantyne
Leica M3 50mm Summicron Agfa CT18
1/500, f4-5.6

Left:
Spot-on timing by the photographer captured this unidentified Class 81 forging north near Carnforth as a local service formed of Class 111 and 105 DMUs crosses overhead on the ex-Midland Railway route to Leeds on 13 September 1975.

Spurred on by developments in other transport modes, BR's planners in the 1960s recognised that electrification of the West Coast main line northwards to Glasgow needed to become a priority if rail was to remain competitive on this profitable route. An initial submission in April 1968 acknowledged not only that electrification could bring major cuts in journey times but also identified the relatively intangible benefits that could be expected, such as cleanliness, reliability and an aura of modernity — in short, the phenomenon now known as the 'sparks effect'. Nevertheless, it was not until 23 February 1970 that the Minister of Transport, Richard Marsh (later to become Chairman of the British Railways Board), authorised the scheme with a view to introducing a full electric timetable in 1974. *Gavin Morrison*
Pentax SP1000 Kodachrome 25

Above:
The BRB submission for electrification to Glasgow included a requirement for additional motive power, and authorisation was forthcoming for 35 new locomotives. The result was Class 87, a GEC design closely based on Class 86 but differing primarily through the use of traction motors mounted on the body frame rather than on the axle. The continuous rating was 5,000hp which, together with multiple-unit control, gave the locomotives an advantage over the gradients of Shap and Beattock. Gearing allowed a cruising speed of 110mph, although this was not required on a regular basis until 1984. This is the classic view of Glasgow Central, as No 87012 makes a triumphant departure with the 17.30 to Birmingham New Street on 25 July 1976. *David M. Cross*

Left:
The reversion of Class 87 to frame-mounted traction motors (GEC 412AZ) and the adoption of innovatory features such as the new BP9 bogie and externally-mounted Flexicoil suspension meant that it was desirable to obtain some service experience before commissioning the new fleet. To this end three 'AL6' locomotives, Nos E3143/50/91, underwent conversion to Class 87 specification at BREL Crewe in 1972, emerging as sub-Class 86/1. They worked interchangeably with other members of Class 86 until electrification through to Glasgow when they took up Anglo-Scottish diagrams. Searing through the Scottish lowlands at Lamington early in the morning of 27 June 1976 is No 86103 (formerly No E3143) with a Euston-Glasgow sleeper. *The late Derek Cross*
Agfa 50S

Right:
Due in part to their Bo-Bo wheel arrangement and light weight, Classes 81-87 are often worked in pairs on heavier freight services where the adhesion provided by a single locomotive is insufficient. Most West Coast main line Freightliner trains run at night, staging at Crewe Basford Hall where a pair of electrics are attached for the run over the northern fells, but some workings run throughout from Willesden, providing the sight of double-heading south of Crewe. One such train is seen here passing Rugby in the care of Class 85s Nos 85008/09 on 25 May 1978. Class 85, like all the early classes, is not equipped for multiple working, so in this instance a driver would have been rostered for each locomotive. *John Cooper-Smith*
Mamiya 645 150mm Sekor Ektachrome 64
1/250, f5.6

Above left:
No 87101 rests at Carlisle while working a Glasgow-Euston service on 3 September 1979.

Experiments with 'AL3' No E3100 in the 1960s revealed the desirability of stepless voltage control in improving the efficiency of locomotive handling, although it was not considered worth persevering with the transductor equipment fitted in that locomotive. By the mid-1960s electronic thyristors had been developed and to provide an in-service testing facility an additional Class 87 (No 87101) was ordered, with thyristor control equipment. It went to the Research Division at the Derby RTC in March 1975 and did not enter revenue-earning service until January 1977. Although the continuous rating of the locomotive, at 4,850hp, was slightly lower than its conventional Class 87 sisters, the ability of the control equipment to match power, and thus tractive effort, to the load meant that under normal conditions its haulage capability was actually superior. The lessons learnt were invaluable and the latest Class 90s can trace their origin directly back to No 87101 which received the name *Stephenson* at Manchester Piccadilly on 12 October 1977. Its nameplates came from No 87001 following the announcement that the '87s' were to be known as the 'Royal Scot' class. *Norman Preedy*
Pentax Spotmatic 50mm Takumar Agfa CT18 1/125, f8

Below left:
Pre-echoing the 'large logo' livery that was to become standard on several main line diesel classes in the early 1980s, Class 86/2s Nos 86214/35 received an early variant of this scheme in 1979 in preparation for the 'Rail 150' celebrations the following year. This picture clearly shows the large yellow panel, incorporating the coat of arms of the Liverpool & Manchester Railway, which was painted on the bodysides of each locomotive. In this 9 May 1982 view, Liverpool Lime Street station plays host to No 86235 *Novelty*.

Sub-Class 86/2 comprised 58 locomotives with GEC 282BZ traction motors which were fitted with SAB resilient wheels and Flexicoil suspension in the 1970s in order to reduce track forces at high speeds. *David Rapson*
Canon AE1 50mm Canon Kodachrome 64 1/30, f6.3

Right:
By the time this photograph of Class 84 No 84003 at Crewe Holding Sidings was taken on 21 March 1978, four of the class had already been withdrawn. This locomotive was one of the last two to be taken out of traffic, in November 1980, and was sent to the Derby RTC to provide spares for No 84009 which was due to undergo conversion to a mobile load bank (see page 41). Stripped of certain components, it languished at Derby for several years, its condition gradually deteriorating, until receiving the merciful attention of the cutter's torch at Vic Berry's of Leicester at the end of January 1986.
Barry Nicolle
Pentax SP2 50mm Takumar Agfa CT21
1/250, f5.6-8

Right:
During the late 1970s the West Coast main line became notorious for inflicting an appalling ride on passengers and, following the creation of sub-Class 86/2 to support Class 87 on high-speed services, the remaining members of Class 86 were restricted to freight and parcels work at a maximum speed of 80mph in an attempt to curb damage to the track. The operators were soon to identify a shortage of 100mph motive power, however, so as an interim measure 20 Class 86/0s were fitted with SAB resilient wheels from 1980 and reclassified '86/3' (altering only the third digit of the number) as a low-cost alternative to the full Class 86/2 specification. The result was a 100mph locomotive that was kinder to the track, although drivers continued to complain of poor riding. This interesting picture shows Class 86/3 No 86326 having undergone wheel conversion but still carrying its Class 86/0 running number. The locomotive never ran in this form, and the correct number was applied soon afterwards. It is seen at Crewe ETD on 1 March 1980 waiting for its pantograph to be fitted. *Barry Nicolle*
Pentax SP2 50mm Takumar Agfa CT18 1/125, f8

Above:
Cast in silhouette by the setting sun, an unidentified Class 86 rushes north along the West Coast main line at Bamfurlong, south of Wigan, with the 19.30 Birmingham-Glasgow on 16 April 1981.
Robert Osborne
Canon AE1 Kodachrome 64

Right:
With the Clyde Valley bathed in the diffused light of early sunrise, an anonymous Class 87 momentarily intrudes while hauling the 23.30 Euston-Glasgow. This magnificent scene was recorded at 05.39 on 1 June 1985. *Antony Guppy*
Canon A1 135mm Canon Kodachrome 64
1/250, f5.6

Above:
Against a backdrop of threatening storm clouds, Class 86/0 No 86016 powers the 11.45 Euston-Aberdeen through Beattock on 2 August 1979, before the 80mph restriction was imposed on this sub-class. Nos 83013 and 86024 wait in the loop with a train of ammonia tanks. *Mrs D. A. Robinson Pentax 6×7 150mm Takumar Ektachrome 200 1/1000, f4.5*

Right:
By the early 1980s the remaining early series ac electrics were confined largely to freight and parcels workings, with forays on to passenger duties being limited mainly to overnight and relief services. The reluctance to entrust the locomotives with Class 1 work stemmed from their high failure rate compared with Classes 86 and 87 rather than from any lack of capability on the road. On a fine 23 October 1982 a sparkling 'roarer', No 81005, rounds the curve at Bessie Ghyll, five miles south of Penrith, with an up relief. *Mrs D. A. Robinson Pentax 6×7 150mm Takumar Ektachrome 200 1/1000, f4*

ELECTRIC
LOCOMOTIVES
MUST NOT
PROCEED TOWARDS
SIDINGS 11 - 23

Left:
Heavy metal line-up at Mossend. With blazing lights lending an air of purpose, representatives of Classes 87, 86/3 and 85 stand at the head of southbound Freightliner trains on 13 September 1985. The multiple working facility of the former two classes makes them especially suitable for such duties. A pair of '87s' is permitted a maximum load of 1,260 tons over the northern fells whereas a single '86/2', a sub-class not equipped for multiple working, can manage only 510 tons, partly on account of its high gearing. *Les Nixon*
Pentax Spotmatic F 135mm Takumar
Kodachrome 64 1min, f8

Above:
Put into store for the second time in the autumn of 1982 and withdrawn in July 1983, Class 83s Nos 83002/5/8/13 stand at the head of the scrap line at Manchester Longsight depot on 19 August 1983, alongside redundant Gloucester Class 100 DMUs. Two former residents of this siding, Nos 83012/15, had already been reinstated for empty coaching stock workings out of Euston, but these remaining locomotives met their demise at Vic Berry's Leicester scrapyard the following year.
Jon Bowden
Canon AE1 50mm Canon Fujichrome 100
4sec, f4

0000
86 206

Left:
Members of the Civil Engineer's Department working on the Down slow line near Bushey keep a respectful distance as Class 86/2 No 86206 hurries past on the Up fast with the 08.59 Wolverhampton-Euston on 9 September 1978. Three months later this locomotive became the second Class 86 to be named, gaining *City of Stoke-on-Trent* plates in a ceremony in December as part of a wider naming policy for the class. *Antony Guppy*
Praktica LLC 50mm Ektachrome 64 1/500, f4

Above:
Essential maintenance of track and overhead line equipment is generally undertaken during weekend engineering possessions, and diversion of electrically-hauled trains regularly results in diesel haulage of electric locomotives over unfamiliar routes. Over the weekend of 29/30 September 1984 West Coast main line services were being routed via Manchester Victoria where Class 87 No 87028 *Lord President* makes a most unusual sight on the evening of the 30th stabled in a siding, presumably due to a brake failure. *John Robinson*
Rolleicord 75mm Xenar Agfa CT18 30sec, f5.6

Above:
Oxenholme, BR's gateway to Lakeland, has changed little since electrification, retention of the overall roof covering the Windermere branch platform helping to retain the station's character. Class 86/2 No 86246 *Royal Anglian Regiment* has just completed the descent from Grayrigg with an up express on 13 September 1986. Class 86s started to receive InterCity livery in July 1984 when No 86242 was outshopped from BREL Crewe in this style, although with a small yellow warning panel, following a general overhaul. By the end of 1986 just over half the class had been so treated.
Les Nixon
Pentax 6×7 105mm Takumar Ektachrome 100 1/500, f5.6

Right:
October 1987 witnessed the roll-out of the first Class 90 locomotive, No 90001, to great admiration from the railway press. The locomotive went to the Engineering Development Unit at the Derby RTC for tests on 31 October, and later began testing on the West Coast main line. It was joined by No 90002 at the end of February 1988 and thereafter members of the class began to infiltrate West Coast main line diagrams to accumulate in-service mileage. Unfortunately some of the admiration evaporated due to problems that became apparent under service conditions, partly caused by shortcomings with relays and related electrical equipment.

Class 90 No 90012 glides round the curve at Greenholme on the ascent to Shap with the 14.30 Euston-Glasgow on 28 March 1989.
Peter J. Robinson
Pentax 6×7 150mm Takumar Fujichrome 400 1/1000, f6.3

INTERCITY
90 012

Left:
During the 1980s Class 85 continued to provide the core motive power for Speedlink services over the northern section of the West Coast main line where loadings and speeds generally decreed that a single locomotive was sufficient. In this beautifully-lit portrait, No 85038 is doubtless having no trouble observing the 60mph maximum speed allowed for Class 6 workings as it sweeps through Low Gill with 6S97, the 08.57 Warrington-Mossend, on 24 September 1987. The two VEAs (second and third wagons behind the locomotive) are carrying explosives, the other three wagons being provided to maintain the required barrier distance between them and the locomotive in front and Cartics behind. During 1989, 10 Class 85s were earmarked for continued use on Railfreight services and reclassified '85/1'. *Peter J. Robinson*
Pentax 6×7 105mm Takumar Ektachrome 200 1/1000, f4

Above:
A lengthy southbound parcels train is wheeled effortlessly downgrade near Mealbank, north of Oxenholme, in the care of Class 81 No 81006 on 11 June 1988. This locomotive was to remain in service for less than four months after this picture was taken, being withdrawn early in October due to the poor condition of its bodywork. *Dave McAlone*
Bronica ETRS 75mm Zenzanon
Ektachrome 100 1/500, f5.6

After a lull of several years BR's main line electrification programme resumed with the authorisation of the Colchester-Norwich scheme in December 1981, although the need to replace much manual signalling equipment meant that electrification work proper did not get underway until 1983. Monday 13 May 1985 saw the start of electric working to Ipswich, and eight Class 86/2 locomotives were based at Ilford for the service, although remaining allocated to Willesden for maintenance. The following two years were a period of operational inconvenience at Ipswich since the station became a traction changeover point, diesel haulage being required for the 46 miles onward to Norwich until May 1987.

The time-honoured location of Belstead Bank provides the setting for Class 86/2 No 86218 *Planet* with a Norwich-Liverpool Street service on 5 August 1986. *Michael J. Collins*
Olympus OM2 Spot 100mm Zuiko
Kodachrome 64 1/500, f4.5

Originally electrified as far as Bishop's Stortford in 1960, the Liverpool Street-Cambridge line was finally wired as far as the university city in time for the start of the 1987 timetable. Network SouthEast, the sponsoring Sector, even went to the extent of painting Class 86/4 No 86401 in its distinctive livery for an inaugural run on 23 March 1987, although the first Class 86 to reach Cambridge was actually No 86204 *City of Carlisle* on a test run on 13 January of that year. On 26 August 1988 it was the turn of No 86242 *James Kennedy GC* to perform Cambridge line duties, and it is seen here passing the Saxby & Farmer-built Great Eastern Railway signalbox at Roydon with the 09.35 from Liverpool Street. *Paul Shannon*
Olympus OM1 100mm Zuiko Kodachrome 64 1/500, f4-5.6

Above:
On 9 May 1985 Class 86/2 No 86241 *Glenfiddich* threads the trackwork on the exit from Manchester Piccadilly with the Up 'Manchester Pullman', just a few days before Mk 3 stock was introduced on this service. *John Whiteley*
Olympus OM1 85mm Zuiko Kodachrome 64 1/500, f4

Right:
Rapidly closing the 193-mile gap between Euston and Liverpool is No 87015 *Howard of Effingham*, passing Tamworth Low Level with the 13.00 service on 9 May 1987. In common with other members of the class, the locomotive is fitted with a Brecknell-Willis 'Hi-speed' pantograph, designed to optimise current collection at 110mph and a prerequisite for regular running to be authorised at that speed. *Chris Milner*
Canon A1 85mm Canon Fujichrome 100 1/500, f4.5

Above:
Although commercial and operating considerations decreed that the ill-fated Advanced Passenger Train (APT) should be unitary in execution, the 4,000hp power cars were, in effect, cabless locomotives; moreover, they provided the conceptual basis behind the development of Class 91 in the late 1980s. The APT was a technological *tour-de-force*, but was simultaneously over-engineered and chronically under-funded, and a series of crippling failures during the week of its service launch to the media in December 1981 dealt a blow to the train's credibility from which it never recovered. In addition, some would say that obsession with the technical aspects, notably the 9° tilt facility, clouded the APT's true purpose, since for a 'train of the future' it featured a cramped, spartan and dated interior, difficult entry and exit for passengers, and a rough ride.

On 26 February 1986, the last two operational units, Nos 370001/03, make a superb sight at Lambrigg while on a Crewe to Carlisle test run. *Peter J. Robinson*
Pentax 6×7 150mm Takumar Ektachrome 200 1/1000, f5.6

Right:
In 1984 it was announced that all Class 86/0 and 86/3 locomotives would be fitted with flexicoil suspension and reclassified '86/4' to provide a standardised 100mph fleet, primarily to handle parcels and secondary Class 1 work. In this wintry scene swirling snow shrouds the late-running 07.10 Manchester Victoria-Glasgow as it emerges from the cutting just north of Shap village in the care of No 86407 on 23 February 1989. *Dave McAlone*
Bronica ETRS 75mm Zenzanon Ektachrome 200 1/500, f5.6

The first ac electric locomotive to receive a Level 5 repair at Stratford Major Depot was Class 86/2 No 86229 *Vulcan Heritage*. The locomotive was photographed from atop the works crane on 3 February 1987, in the course of being stripped down. Following completion of the work, No 86229 was outshopped in an experimental version of the InterCity livery with the upper bodyside panel being painted in a dark shade of brown. *Brian Morrison*
Nikon FE 35mm Nikkor Kodachrome 64 1/30, f2

A ride in the cab of an ac electric is an experience not quickly forgotten, and it is easy to underestimate the skill required to drive such a locomotive smoothly. Not only does the driver have to master the principle of applying power by tapping current through 'notching up' and reducing power by 'notching down' (in a non-thyristor locomotive), but he must keep a careful eye on the traction motor ammeters to ensure that the motors are not overloaded. Also, the driver must be aware of the location of 'neutral sections', even at night, in addition to his normal route learning requirements. Since all the ac electric classes are capable of well over 100mph, service maxima notwithstanding, there is, furthermore, a very real possibility of overspeeding if a watchful eye is not kept on the speedometer! The photograph was taken, with BR permission, in the cab of a Class 86/2 on the Great Eastern main line. *Brian Morrison*
Nikon FE 24mm Nikkor Kodachrome 64
1/30, f16 (balanced flash)

Left:
By the early 1980s schedules on the Euston-Glasgow route had slipped somewhat and BR's market share was diminishing. In an attempt to counteract this Intercity introduced 110mph running on certain Anglo-Scottish services in May 1984, using Class 87s fitted with Brecknell-Willis pantographs. No 87006 *City of Glasgow* was painted in an experimental grey version of the 'large logo' livery for the service, although this proved to be short lived, losing out to the 'executive' scheme as the standard InterCity livery. Unfortunately 'Inter-City 110' was not a commercial success on the Glasgow route, although 110mph running was introduced on certain Pullman services to Liverpool and Manchester in 1986. No 87009 *City of Birmingham* prepares to lift an Up service out of Liverpool Lime Street on 14 June 1986.
Gavin Morrison
Pentax K1000 50mm Pentax Kodachrome 25

Above:
September 1960 marked completion of electrification work between Manchester Piccadilly and Crewe. At that time the Beeching Report, with all its implications for route and track rationalisation, was still three years off, and electrification did not mark the wholesale lifting of little-used track that accompanied later schemes. Stockport still presents a jumble of trackwork as No 86245 *Dudley Castle* arrives at the station with the 13.20 Manchester-Birmingham on 1 April 1989. *Paul Shannon*
Olympus OM1 135mm Zuiko Kodachrome 64 1/250, f5.6

Above:
With dawn just beginning to break, Class 87 No 87030 *Black Douglas* pauses at Platform 6 at Crewe with the 02.25 Preston-Birmingham on 18 June 1986. At this time just under half the class was in standard blue livery. *Chris Shaw*
Pentax ME Super 50mm Pentax Kodachrome 64 30sec, f5.6

Right:
An image of the modern railway as No 86215 *Joseph Chamberlain* clicks away to itself amid the rebuilding work at Liverpool Street at the head of the 21.30 to Norwich on 12 April 1989. This unusual noise made by an ac electric at rest is caused by the cooling of electrical equipment. *Chris Shaw*
Pentax LX 28mm Pentax Kodachrome 25 2½min, f8

15

Left:
Skirting the still waters of the Grand Union Canal at Northchurch is Class 86/2 No 86248 on 6 December 1986. The standard class stock comprises one of several sets of declassified Mk 2f FOs that were in use on West Coast main line daytime services during this period, partly to compensate for coaches lost in the Colwich crash the previous September, an incident which also resulted in the destruction of two Class 86s, Nos 86211/429. *Michael J. Collins*
Olympus OM2 Spot 50mm Zuiko
Kodachrome 64 1/500, f4

Above:
On a bright but bitterly cold 10 December 1988, No 90010 hurries through rural Hertfordshire near Berkhamsted forming a northbound light engine movement, believed to be a driver training run. At this time the '90s' were becoming quite a common sight on the West Coast main line, the end of the previous month having seen six passenger diagrams allocated to the class. *Chris Shaw*
Pentax LX 50mm Pentax Ektachrome 100
1/1000, f4

A scene of activity at Northampton on 4 August 1989 as No 86207 *City of Lichfield* stables the stock of an evening commuter service from Euston while Network South East-liveried No 86401 *Northampton Town* runs round another 'cobbler' set.

Introduction of Class 321 EMUs spelled the end of these locomotive-hauled services in the spring of 1990, and No 86401 has since been allocated to the Parcels Sector based, in common with other non-InterCity electrics, at Crewe ETD.

Antony Guppy
Canon A1 50mm Canon Kodachrome 64
1/500, f4

The year 1987 saw BR holding major celebrations to mark 21 years of its use of the InterCity brand name, despite the Sector of that name which organised the festivities not yet having reached its sixth birthday! On 1 May a re-run of the 1966 'Merseyside Express' from Euston to Liverpool was staged, comprising Class 86/4 No 86426 hauling a rake of maroon Mk 1 coaches. Since none of the '86s' was fitted with multiple working equipment in 1966 it is surprising that an '86/2' was not selected, but in the event No 86426, complete with cab-end paraphernalia, was outshopped in an approximation of its 1960s condition. Safety considerations dictated that a full yellow warning panel was necessary but there was no such excuse for the application of a coat of rail blue that was clearly too dark compared with the 1960s shade. The locomotive nevertheless made a fine sight. Here it is seen on more mundane work, heading a north-bound train of steel coil empties near Scout Green on 3 November 1988. *Chris Shaw*
Pentax LX 50mm Pentax Kodachrome 64 1/1000, f2.8

Late in 1988 Freightliner became part of Railfreight Distribution, and the sub-Sector lost no time in converting eight Class 86/2 locomotives to become a new sub-Class, '86/5', to handle its services over the gradients of the West Coast main line. The original traction motors were retained but were regeared, the maximum speed being limited to 75mph in consequence. Multiple working equipment was to have been fitted at a later date. Cynics were not surprised when, by the summer of 1989, InterCity had identified a need for more high speed passenger locomotives and the decision was taken to re-convert the '86/5s' to their former classification. By November all had reverted to their previous identities.

Resplendent in Railfreight livery, but minus sub-Sector symbols, No 86503 *City of Lancaster* (formerly No 86205) traverses the heart of Manchester between Oxford Road and Piccadilly stations with 4L76, the 13.16 (SO) Trafford Park-Tilbury on 6 May 1989. Since 15 May 1989 passenger service frequency over this stretch of line has been such that only two paths an hour exist for freight services. *Paul Shannon*
Olympus OM1 50mm Zuiko Kodachrome 64
1/250, f5.6-8

The North London link between the West Coast and East Coast main lines was electrified with Railfreight sponsorship during 1987 to permit inter-Regional electric haulage of the Sector's trains. The first revenue-earning electrically hauled freight ran on 18 April 1988, but until May 1989 the only services to benefit from regular electric haulage were the Dagenham-Halewood automotive trains. From May, however, three Freightliner diagrams, involving seven trains, were booked for electric locomotives. One of these services, 4M47, the 10.18 Felixstowe-Willesden (electrically hauled from Ipswich), passes Highbury & Islington in the care of Nos 86620 and 86416 on 3 July 1989. Railfreight Sector Class 86/6 No 86620, regeared for 75mph and fitted with phosphorous brake blocks, was carrying its fifth number since introduction as No E3114 in 1965, having previously been classified 'AL6', '86/0', '86/3', and '86/4'. *Paul Shannon*

Olympus OM1 100mm Zuiko Kodachrome 64 1/250, f5.6-8

Above:
Class 86/2 No 86236 *Josiah Wedgwood* had only just received TDM equipment for push-pull working with Mk 3 DVTs when this photograph was taken at Soho Junction, Birmingham, on 15 August 1989. The train is the 15.25 Wolverhampton-Euston, and the DVT is being employed as a standard BG on this occasion. *Peter Tandy*
Mamiya 645 1000S 150mm Sekor
Fujichrome 100 1/500, f4.5

Right:
Once the Class 86/6 conversion programme is complete sub-Class 86/4 will number just seven locomotives allocated to the Parcels Sector at Crewe ETD. In common with the Sector's Class 90s, the '86/4s' will receive red and grey livery. No 86419 *Post Haste*, the first locomotive to be so treated, stands in the sunshine at Stratford on 13 July 1990. *Brian Morrison*
Nikon FE2 50mm Nikkor Ektachrome 64
1/125, f8

86 419
Post Haste

Despite its nominal allocation to individual business Sectors the majority of the London Midland Region's ac electric fleet is diagrammed for use on more than one Sector's services. For this reason Railfreight was denied its wish to see its first batch of Class 90s painted in the smart new two-tone grey Railfreight livery, the BR main line scheme being adopted instead. Curiously this state of affairs was later partly reversed, and, following an acceptance delay of several months, No 90037 and subsequent Railfreight Class 90s entered service in the Sector's house colours.

An unidentified Railfreight Class 90 crosses the River Nene at Northampton with the diverted 15.32 Blackpool North-Euston on 27 May 1990.
Antony Guppy
Canon A1 50mm Canon Kodachrome 64
1/500, f4

Diverted via Hertford North due to engineering work an unidentified Class 91 rejoins the main line at Langley Junction with the 16.40 King's Cross-Leeds on 28 May 1989. With an IC125 power car bringing up the rear and providing on-train power supply a total of 8,350hp (continuous) is available, although such is the power of the '91s' it is highly unlikely that the extra diesel power has any discernible effect on performance. *Paul Shannon Olympus OM1 135mm Zuiko Kodachrome 64 1/250, f5.6-8*

The setting sun highlights the blank faced coaches of the London Midland Region's premier train, the Up 'West Coast Postal', reporting number 1M44, as it glides smoothly south near Carlisle in the care of No 86421 *The London School of Economics* on 15 July 1988. *Dave McAlone*
Bronica ETRS 150mm Zenzanon
Ektachrome 200 1/125, f5.6